THIS BOOK IS DEDICATED TO VICTORIA AND ROBERT ZOELLNER

ROLEX

All Aboard!

A Tour of the Holiday Train Show at

THE NEW YORK BOTANICAL GARDEN

PHOTOGRAPHS BY JOHN PEDEN

THE HOLIDAY TRAIN SHOW at The New York Botanical Garden has become such an eagerly anticipated New York City tradition that many people think the Garden has offered this wonderful exhibition for decades. Actually, the show is young: 1991 was the first year.

Each year more and more visitors flock to the Garden to enjoy the historic houses, trains, lights, and music. They find holiday cheer in the form of New York landmarks creatively interpreted in bark, branch, leaf, and seed pod rather than the original brick and mortar, and set among hills, plantings, waterfalls, gorges, and winding creeks. The imagination and craftsmanship of show creator Paul Busse combine to form a fantastical botanical cityscape traversed by large-gauge trains in a lush greenhouse world.

For many, the show evokes cherished memories from childhood, real or imaginary, centered on trains. For years visitors have brought their children here and may now be bringing their grandchildren to this magical 6,000-square-foot exhibition set in the glorious Enid A. Haupt Conservatory, named in honor of the Garden friend and philanthropist whose generous support makes exhibitions like the *Holiday Train Show* possible.

When Victoria and Robert Zoellner first saw the show more than a decade ago, they imagined it growing into an enchanting annual holiday tradition. They realized this vision; their steadfast generosity has helped build this exhibition into a favorite holiday outing for families across the New York metropolitan region.

I hope you enjoy this book as a photographic memento of the show and that you bring family and friends to the *Holiday Train Show* for years to come.

Gregory Long

Gregory Long, *President*
The New York Botanical Garden

NORTH POLE SNOWFLAKE RAIL ROAD
Amtra

When you visit the *Holiday Train Show* in the Enid A. Haupt Conservatory, you enter a magical scene. Everywhere you look, there is an enchanting view of New York landmarks, which, amazingly enough, are made from bark and twigs rather than from brick and steel. In this fantastical world, miniature steam engines and sleek passenger trains glide past a diminutive Statue of Liberty dressed in palm leaves and a knee-high Apollo Theater with a marquee lettered in radish and catalpa seeds. Trestles and bridges are made of willow and grapevine and tunnels of hollow logs and twisted hazel. Tucked among these landmarks of dried plants are beautiful, thriving specimens of anthuriums, poinsettias, cacti, and conifers. Before long, you begin to wonder who lives in this charming world and who travels on the trains just coming into view. The wonder and excitement of model trains, meticulous artisanship, and garden features all come together in the *Holiday Train Show,* one of New York's most popular holiday traditions.

A festive glow and the season's trimmings welcome you along the path to the Conservatory. A 25-foot evergreen sparkles with brilliant colors and botanical ornaments. Snowflake-shaped lights twinkle along the allée of honey locust trees. At dusk, this tranquil winter scene invites reflection, apart from the City's holiday hustle. Upon entering the Conservatory's majestic arched doorway, you are drawn into a holiday adventure of trains, trees, music, and lights inside the vast ribbed frame of graceful steel latticework and curved glass. But as you come in from the cold, you discover so much more—a miniature New York artfully crafted of plant material and the lush flora and warm feelings of a tropical rain forest in the midst of a New York City winter.

The New York Botanical Garden and show creator Paul Busse began their collaboration in 1991, and every year since the *Holiday Train Show* has become more elaborate. It takes about 10 days to lay out the 1,200 feet of track, get the trains running, construct the trestles, position the buildings, tuck in the hundreds of plants, and place the lights just so. The goal of the show is to create an emotional response in the spectators, to draw you into the fantasy of this miniature world. Never do you feel like a giant looking down upon a lilliputian figure-eight train track, where all the trains and villages are visible at once. The track arrangement consists of simple, unconnected loops, with the tracks woven into the landscape and screened by houses and plants. With as many as 11 trains running at a time, there is always that delightful feeling of anticipation, of waiting for a train to reappear.

While awaiting its arrival, your focus shifts to the remarkable botanical creations that are the crown jewels of the exhibition. Some are well-known landmarks, others lesser known or even buildings that are no longer in existence. Working from photographs, Paul and his crew create easily recognizable small-scale translations rather than precise replicas. Just as a skilled caricaturist captures a personality with a few strokes of a pen, they capture the essential,

distinguishing elements of a building with plant materials. Using bark, berries, seeds, and vines adds unusual texture, intricate detail, and charm and, of course, is the key to making the whole design utterly harmonious in the garden. Paul collects interesting natural material year-round. He admits to not being able to walk through the woods in his native Kentucky without bringing something back, whether it is a hollow log for a tunnel or textured bark for a roof.

The *Holiday Train Show* combines the designer's love of trains with a landscape architect's eye for detail and an artisan's imaginative use of plants. Together with the impressive venue of the grand Victorian Conservatory, the show makes for the perfect holiday experience. Now in its 12th year, the *Holiday Train Show* is one of New York City's greatest holiday traditions, and according to the *New York Times,* "one of New York City's best holiday gifts to itself."

About the Trains

The trains in The New York Botanical Garden *Holiday Train Show* are G-gauge trains. Designed to be run outdoors, they have motors and gears that are housed in weatherproof casings. They are powered with low voltage on gauge-1 brass track, which measures 1 3/4 inches (45 mm) between the rails. A large-scale steam locomotive can measure over 2 feet long and weigh as much as 10 pounds.

The locomotives and railroad cars in the *Holiday Train Show* represent American trains from late-1800s steam engines to today's most modern high-speed passenger trains. Model trains, however, have existed since the inception of the railroad industry in the 1830s. Many of the earliest models were built in response to competitions sponsored by railroads to encourage inventors to solve early engineering problems. Others were built to secure the public's interest or to entice financial backers to support new railway lines. Afterwards, these steam models—some of which

were large enough to ride on—were often given to the investors' children and installed outdoors on spacious estates.

Soon toy manufacturers started making model trains. But just as in the world of full-size trains, there was no consistency in the gauge of the track or the scale of the trains. This problem was solved in the 1880s when the German company Märklin established standardized gauges and scales. The first standardized gauges were called 1, 2, and 3. Gauge 1, at 45 mm, was the smallest and gauge 3, at 62 mm, was the largest. Before long, almost all American and European model manufacturers had adopted these standards. Most of these railroads were still built for use outdoors and eventually the term "garden railway" came to be used to describe model railroads which, with suitable landscaping, became decorative and entertaining features in the garden.

It wasn't long before toy manufacturers realized that they were missing a market—those excluded from the hobby because they had no garden—and they began making smaller trains. Gauge 0 was 32 mm between the rails and quickly became very popular. These trains were intended to be run indoors and were inexpensive. The engines were powered either by wind-up clockwork, steam, or electricity. In the United States, Lionel was the dominant manufacturer and focused almost exclusively on these smaller models. As a result, although outdoor garden railways continued to be popular in Europe, few people in the United States were familiar with them.

In 1967 the German firm of Ernst Paul Lehmann introduced a new concept at the Nuremberg Toy Fair: Lehmann's Gross Bahn (LGB), or Lehmann's Big Trains, were made of modern UV resistant plastic and were modeled on German narrow-gauge trains.

They were sturdy and designed to run outdoors on standard gauge-1 track, thus they were built at the unusual scale of 1:22.5. These new LGBs were not immediately popular in the United States as there were no models of American trains. However, in 1985, LGB introduced a model of a real Colorado narrow-gauge steam engine, the Mogul. American narrow gauge railroads were designed as short line systems where the terrain was too steep for standard railroads. With models of American trains now available, garden railroading as a hobby took off in the United States. Nowadays, many U.S. manufacturers make large-scale trains—among them Aristo-Craft, Bachmann, Lionel, and USA Trains.

About The New York Botanical Garden

The New York Botanical Garden is a museum of plants alive with discovery, from the landmark Enid A. Haupt Conservatory offering an ecotour of the world under glass, to the innovative Everett Children's Adventure Garden where children explore nature and plant science, to 48 magnificent gardens and plant collections on a 250-acre historic site.

The Garden is also a global leader in botanical research. Its International Plant Science Center combines cutting-edge technology with more than 100 years of scientific research to document and preserve the plant biodiversity of the Earth. It is considered one of the most intensive and distinguished botanical research programs in the world.

In 1891 an eminent Columbia University botanist, Nathaniel Lord Britton, and his wife, Elizabeth, also a botanist, inspired by their visit to the Royal Botanic Gardens, Kew, near London, determined that New York should possess a great botanical garden. A magnificent site was selected in the northern section of the Bronx, part of which had belonged to the vast estate of Pierre Lorillard, a leading tobacco merchant. The land was set aside by the State Legislature for the creation of "a public botanic garden of the highest class" for the City of New York. Prominent civic leaders and financiers, including Andrew Carnegie, Cornelius Vanderbilt, and J. Pierpont Morgan, agreed to match the City's commitment to finance the buildings and improvements, initiating a public-private partnership that continues today.

The historic Enid A. Haupt Conservatory, originally constructed in 1902, re-opened in May 1997 following a four-year restoration and construction of *A World of Plants,* an exhibition that includes tropical rain forests; desert galleries; and the *Palms of the Americas Gallery,* the world's most comprehensive collection of palms under glass. The *Holiday Train Show* is just one of the many seasonal exhibitions that take place each year inside the Conservatory. Other horticultural showcases include magnificent displays of spring flowers, dazzling tropicals in summer and fall, and a celebration of orchids in winter.

Flatiron Building Metropolitan Life Insurance Tower Municipal Building
New York Public Library Radio City Music Hall

Chrysler Building

Empire State Building

Grand Central Terminal

Solomon R. Guggenheim Museum

New York Public Library (1911)
Fifth Avenue between 40th and 42nd Streets, Manhattan
Considered one of the finest Beaux-Arts-style buildings in the United States, the Library's first Director, Dr. John S. Billings, came up with the basic design that was then fully developed by the architects Carrère & Hastings. The marble lions that flank the front steps were renamed by Mayor LaGuardia for the qualities he felt New Yorkers would need to survive the Depression: Patience and Fortitude. Our lions are made from pear-shaped pods with grapevine tendril tails, okra seed eyes, and wild meadow grass manes. The library's stone work is made of bark and the central roof is made of magnolia leaves.

Grand Central Terminal (1903–1913)
42nd Street and Park Avenue, Manhattan
This Beaux-Arts-style train station embodies the romance and glamour of the Golden Age of Rail and remains a civic design triumph of architecture, technology, and city planning. The building is topped by a 1,500-ton sculpture of Mercury, Hercules, Minerva, and an eagle, symbolizing commerce on the move, sustained by moral and mental energy. Our façade is made of sand and blocks of sycamore bark and roofed with southern magnolia leaves. The lampposts are fashioned from honeysuckle stems and acorns. The rooftop sculpture is crafted from dried strawflowers, arborvitae, and yarrow.

New York Public Library

Solomon R. Guggenheim Museum (1956–1959)
1071 Fifth Avenue, Manhattan
The Guggenheim, Frank Lloyd Wright's only design in New York City, is one of the city's most noteworthy and controversial buildings. A symphony of triangles, ovals, arcs, circles, and squares, its central spiral form recalls a nautilus shell with continuous spaces flowing freely into one another. The modern and contemporary art is displayed on the walls of a ramp in the center of the spiral that gradually rises to 92 feet, with galleries extending off of it. Our version is made of several large black locust shelf fungi.

Radio City Music Hall (1932)
1260 Sixth Avenue, Manhattan
Considered one of the most important examples of Art Deco style, Radio City Music Hall was designed by Edward Durrell Stone, who later created The Museum of Modern Art. When it opened in 1932, it was the nation's largest hall. The marquee, a full city block long, is reproduced here with willow, pear-shaped pods, and tinted radish and catalpa seeds. The façade is maple bark with doors of red twig dogwood and grapevine tendrils. The window frames are dyed lemon branches and dyed lemon leaves appear between the windows to recreate the Art Deco detail.

Previous spread
Metropolitan Life Insurance Tower (1907–1909)
1 Madison Avenue, Manhattan
Designed by Napoleon Le Brun & Sons, the campanile-style tower of the Metropolitan Life Insurance Company with its slender shaft, enormous roof, and cupola evokes the bell tower of San Marco in Venice. For a time this was the tallest building in the world. Our replica is crafted from honeysuckle, pine cones, corkscrew hazel, and bamboo.

GUGGENHEIM MUSEUM
RADIO CITY
RADIO CITY
Music Hall
RADIO CITY

Plaza Hotel (1905–1907)
Fifth Avenue and Central Park South, Manhattan
Designed by Henry J. Hardenbergh, who also designed the Dakota Apartments, this French Renaissance-style building with its slate mansard roof, gables, and crest is one of New York's oldest hotels. Since the opening of its doors, it has been a favorite choice for balls and weddings, as well as a home away from home for visiting monarchs, magnates, and a little girl named Eloise. Our miniature version is decorated with canella berries, bamboo pieces, corkscrew hazel, and grapevine tendrils.

Plaza Hotel

Flatiron Building (1901–1903)
175 Fifth Avenue, Manhattan
This unusual building, designed by Daniel H. Burnham & Co., marks the beginning of the skyscraper era in New York City. The triangular shape is the result of Broadway's diagonal route through Manhattan's street grid. The rounded apex at 23rd street measures only six feet across. The rounded corners of our version are made of white pine cones, lotus pods, and dried beans. Lemon and oak leaves are used for horizontal ornamentation, and the spire is a pomegranate with a honey locust thorn.

Municipal Building (1907–1914)
1 Centre Street, Manhattan
Designed by McKim, Mead & White, the building straddles Chambers Street and has a tower that lends it an Imperial Roman feel. Our version includes Adolph A. Weinman's sculpture, *Civic Fame,* which rises above the rooftop lantern, reproduced here with cattail stems, sunflower seeds and stems, shelf fungus, and acorns. Bamboo, acorn caps, basket reed, and lotus pods and seeds are used for the rest of the tower.

Chrysler Building (1930)
405 Lexington Avenue, Manhattan
This Art Deco crown jewel of the New York skyline was, for a few months, the tallest building in the world until the completion of the Empire State Building. It juxtaposes linear curves with sharp angles. Our version has a ginkgo leaf surface with corkscrew hazel branches making up the contoured areas. Sugar pine cone scales form the roof, with small gourds projecting from the corners representing the building's eight stainless steel eagle-like gargoyles. The spire is a honey locust thorn.

Empire State Building (1929–1931)
350 Fifth Avenue, Manhattan
At 103 floors it was the world's tallest building for over 40 years until the World Trade Center was built in 1973. Its design is based on the perpendiular style of a simple pencil. World famous for its role in the 1933 movie *King Kong,* in 1945 it withstood a B-25 bomber crashing into the 79th floor in dense fog. Our version's façade is adorned with a lotus pod, hemlock cones, walnut shells, acorn caps, honeysuckle twigs, and sand in grout for masonry.

Flatiron Building Municipal Building Chrysler Building Grand Central Terminal Empire State Building

Grant's Tomb (1891–1897)
General Grant National Memorial
Riverside Drive at 122nd Street, Manhattan
Designed by John H. Duncan, the memorial is a loose copy of Mausoleus' tomb (350 B.C.)—one of the Seven Wonders of the Ancient World—in what is now southwest Turkey. It is the largest mausoleum in North America. New York City was chosen as the final resting place of Ulysses S. Grant in part so his widow Julia could visit daily. In 1902 she was also interred here. The mausoleum's granite and marble are represented here by willow, pine cones, moneta leaves, canella berries, and corkscrew hazel.

Saks Fifth Avenue (1924)
611 Fifth Avenue, Manhattan
Architects Starrett & Van Vleck designed this luxurious emporium in 1922 in a Renaissance-inspired style appropriate to the exclusive character of Fifth Avenue. Our version has balcony rails and posts made of Virginia creeper and screwbean mesquite. The pillars on the fourth floor between the windows are made with sugar pine cone scales, yew bark, and acorn caps. The top floor railings are sugar pine cone scales, eucalyptus pods, and bittersweet vines, while the cornice is crowned with willow, Russian sage, and white pine cone scales. The doors have juniper berry handles.

Metropolitan Museum of Art (1895–1902)
Fifth Avenue between 80th and 84th Streets, Manhattan
Housing one of the most comprehensive art collections in the world, this building is a rich melange of styles. The early Gothic-style structure by Calvert Vaux was absorbed into the Fifth Avenue Beaux-Arts façade designed by Richard Morris Hunt. McKim, Meade & White designed the side wings along Fifth Avenue that were added from 1904 to 1926. Our version is made of dried zinnias, grass fiber rope, arborvitae cones, walnut shells, canella berries, eucalyptus leaves, cinnamon sticks, and various pods.

Frick Collection (1913–1914)
1 East 70th Street, Manhattan
Designed by Carrère & Hastings, this Louis XV-style mansion was built at enormous cost by Henry Clay Frick as both his residence and a home for his art collection.
The design incorporated his plan to leave both to the public. The collection includes more than 1,000 works of art from the Renaissance to the late 19th century. Corkscrew hazel, gourds, and leaves were used to create our mansion. Sweetgum seed capsules, eucalyptus pods, grapevine tendrils, pine cone scales, and acorn caps form the railing.

Gothic Arch (Bridge No. 28)

The Dairy

Central Park, Manhattan

Previous spread, this page, right and overleaf

Gothic Arch (Bridge No. 28)

Gothic Arch (Bridge No. 28) (1861)
Designed by Calvert Vaux and Jacob Wrey Mould, this wonderful Gothic Revival bridge was made of cast iron and steel by Cornell Iron Works. The ubiquity of horses throughout the city when it was built influenced its design and the creation of trails throughout the park, which were deliberately created with curves to discourage horse racing. Our bridge is fashioned of corkscrew hazel, cinnamon slices, canella berries, grass fiber rope, and acorn caps.

Belvedere Castle (1869)
Designed by Frederick Law Olmsted and Calvert Vaux as a Victorian "folly," the turreted castle includes Gothic, Romanesque, Chinese, Moorish, and Egyptian motifs. At the highest natural elevation in the Park, the Belvedere — Italian for "beautiful view" — offers visitors just that. The U.S. Weather Bureau has been collecting data here since 1919. The front façade of our model is brown-dyed flax grass with bark "brickwork." The turrets are made of pear-shaped pods, grapevine, and corkscrew hazel trimming with roof shingles of eucalyptus.

Music Pavilion (1862)
No longer in existence, this replica of the original Victorian-styled cast-iron bandstand was designed by Jacob Wrey Mould and was demolished in 1922 and replaced by the Naumburg Bandshell. Our bandstand base is made from cattails, canella berries, grapevine, pine cone scales, cut gourds, dried wild flax grass, and lotus seeds with columns of redbud and acorn caps. The elaborate roof is fashioned from dried sea grape leaves, small gourds, grapevine, various pine cone scales, and canella berries.

Angel of the Waters (1873)
Bethesda Fountain
Angel of the Waters is the graceful sculpture atop the Bethesda Fountain. The lily in her hand represents Purity while the four figures below represent Peace, Health, Purity, and Temperance. Designer Emma Stebbins, the first woman to receive a sculptural commission in New York City, likened the healing powers of the angel to that of the clean and pure Croton Reservoir water. Our replica is composed of dried orange slices, pear-shaped pods, lemon and eucalyptus leaves, sugar pine cone scales, hemlock cones, and wisteria pods.

The Dairy (1870)
The Gothic Revival design by Calvert Vaux is best described as Swiss chalet meets Gothic country church. The building was intended as a place where children could enjoy a glass of fresh milk, which was not always easy to get in mid-19th-century New York. The roof on our Dairy is composed of eucalyptus leaves with shingles of pine cone scales. The "gingerbread" trim is made of cinnamon stick slices, redbud, canella berries, and acorn caps.

Cop Cot (ca. 1860)
Scottish for "little house on the crest of the hill," the Cop Cot rests on a large outcrop overlooking the pond and is Central Park's largest rustic wooden structure. Our replica is fashioned out of corkscrew hazel, pine bark, wood bark fungus, beech seed husks, yucca seed pods, acorn caps, and pine cone scales.

Angel of the Waters

Cop Cot

Music Pavillion

Snuff Mill (1840)
The New York Botanical Garden, Bronx
The Snuff Mill, the oldest building at The New York Botanical Garden, was part of the vast estate of Pierre Lorillard, a prominent tobacco merchant. Constructed of local fieldstone, the mill used the adjacent Bronx River water power to produce snuff, a tobacco product popular in the 19th century. Our Snuff Mill is sided with tobacco leaves and its roof is covered with strips of natural cork and moss. The ribs of the mill wheel are oak bark covered in woodland moss and edged in bittersweet vine.

Gracie Mansion (1799–1804)
Carl Schurz Park, Manhattan
This Federal-style building, near the junction of the Harlem and East Rivers, has been the official residence of the Mayor of New York since 1942. When built by Scottish shipping magnate Archibald Gracie, it was considered a remote country house with expansive views of New Jersey and Long Island. The railings and porch columns of our version are corkscrew hazel with arborvitae cones and eucalpytus pods as decoration. The main roof is made of palm sheaths, while the porch roof is layered in sphagnum moss, honeysuckle, acorn caps, and corkscrew hazel.

Schumann's Sons

Apollo Theater

171 Duane Street

Schumann's Sons (1909)
716 Fifth Avenue, Manhattan
This fanciful art nouveau building, no longer in existence, once housed a jewelry store on Fifth Avenue. It featured a sinuous glass marquee, a curved roof façade accented on either side by matching urns and an elegant balcony. Our balcony is created from grapevine tendrils and corkscrew hazel. Pine cones, dried pomegranate halves, oak leaves, and various pods and seeds make up the façade. The curtains in the windows are made of Queen Anne's lace seed-heads set in resin.

Apollo Theater (1913–1914)
253 West 125th Street, Manhattan
Orginally Hurtig & Seamon's New Burlesque Theater, it was renamed the Apollo in 1934 and became the Harlem showplace for African-American entertainers. It launched the careers of Billie Holiday and Ella Fitzgerald, among others. Throughout the 1930s and '40s live broadcasts featuring Duke Ellington and Count Basie cemented its reputation as one of the nation's top stages. The marquee and signs of our version use tinted radish and catalpa seeds for lettering with backgrounds and borders of white sand and willow sticks. The façade detail is done with cut cinnamon sticks, canella berries, poppy seed capsule tops, and a variety of seeds.

171 Duane Street (ca.1859)
Manhattan
Commercial buildings in pre-Civil War New York often had cast-iron fronts. They were popular because of the speed and economy with which they could be constucted. 171 Duane Street was originally a three-story brick Federal-style house that was transformed into a commercial building by the addition of two stories and an iron front. Sweetgum seed capsules, nigella seed pods, pear-shaped pods, and various twigs and bark decorate the front of our version.

Washington Arch

Aguilar Library

Edith and Ernesto G. Fabbri House | Carriage House | Lycée Français de New York

Previous spread

Washington Arch (1889–1895)
Washington Square Park, Manhattan
Standing at the entrance to Washington Square Park is the Washington Arch, designed by McKim, Mead & White. In the aftermath of the Revolutionary War, the park was used as a potter's field and dueling ground. In the 1820s, the elite of New York began building elegant town houses around the park, which served as the nucleus of New York's old society, written about so poignantly by Henry James in *Washington Square* (1881). Our arch is crafted of willow, canella berries, walnut shells, and pine bark.

Aguilar Library (1898–1899)
174 East 110th Street, Manhattan
Our model recreates the original building designed by Herts & Tallant, which was expanded and given a new façade in 1905. The library, founded in 1886 as an independent institution to serve immigrant Jews, was named after Grace Aguilar, an English novelist of Sephardic descent. It is now part of the New York Public Library system. The stone work on our replica is made of bark squares. The balcony is formed of pear-shaped pods and corkscrew hazel.

Edith and Ernesto G. Fabbri House (1898–1900)
11 East 62nd Street, Manhattan
Continuing the Vanderbilt family tradition of building elaborate houses for their children, William H. Vanderbilt's daughter Margaret built this beautiful Beaux-Arts building for her daughter and son-in-law, Edith and Ernesto Fabbri. It is now the residence of the permanent representative of Japan to the United Nations. Our version is fashioned from pear-shaped pods, maple leaves, wisteria pods, dried orange slices, screwbean mesquite pods, rice flowers, sugar pine cone scales, acorn caps, and eucalyptus pods.

Carriage House (1909)
124 East 19th Street, Manhattan
Private carriage houses for horses and conveyances were typically located on another, less fashionable block than the residence to which they belonged. This one was eventually converted into a residence itself. Our version's façade is made of sand in grout with a magnolia leaf for the door. The trim and windows are made of bamboo and the window boxes hold dried baby's breath.

Washington Arch

Folies Bérgère Theater (1911)
210 West 46th Street, Manhattan
This exotic building, no longer in existence, was one of the first dinner theaters in New York. The front was tapestried in a Moorish pattern of ivory, gold, and turquoise terra-cotta. In 1955 the theater was renamed the Helen Hayes Theater, but in 1982 it was demolished and Broadway's Little Theater was renamed in her honor. Our replica is made of magnolia and poplar leaves, bamboo, wisteria pods, cattails, sliced cinnamon sticks, beechnuts, star-anise, eucalyptus, pear-shaped pods, and acorn caps.

Lycée Français de New York

Lycée Français de New York (1898–1899)

7 East 72nd Street, Manhattan

Originally the home of Oliver Gould and Mary Brewster Jennings, the curved copper and slate mansard roof, deep-set French windows, and elegant ironwork of this ornate building evoke the opulence of Napoleon III's Paris. This version is made of sand in tinted grout, oak leaves, grapevine tendrils, pine cones, cinnamon stick slices, and dried orange slices. Another version is pictured on page 35.

Town House (ca. 1890)

Manhattan

Originally the term "town house" was used to distinguish a city residence from a country house of the same owner. It is now generally used to describe a home in an urban configuration, such as a rowhouse. "Town house," however, sounding more elegant, is generally better for sales. Some of New York's 19th- and early 20th-century town houses were very elegant indeed. Our town house's façade is sand in grout with pilasters made of bark strips and twigs. The capitals are acorn caps and canella seeds. The balcony is made of shelf fungus, grapevine tendrils, and moss.

Mrs. George T. Bliss House (1907)

9 East 68th Street, Manhattan

Architecturally, this elegant house is similar in spirit to that of the famous British Regency architect Sir John Soane. The Metropolitan Museum of Art now possesses the large, stained glass window designed by John La Farge that was once a prominent feature of the house. The window depicts a classically garbed allegory of Welcome modeled on Mrs. Bliss's daughter. Our version's façade is sand in tinted grout with birch branches for the half columns. The column bases and lintels are made of bark strips. Grapevine tendrils and eucalyptus pods are used for ornamentation.

Lycée Français de New York

Town House

Mrs. George T. Bliss House

Queens Insurance Company

David S. Brown Store

Queens Insurance Company (1877)
37-39 Wall Street, Manhattan
No longer in existence, this red brick and stone structure was designed in a Venetian-influenced late Romanesque/early Gothic style. It was distinguished by its over-scaled stoop and entrance portico and by the expansive glazed arch storefront at street level. Our storefront windows are configured from two different kinds of bark, while the sand in tinted grout façade is decorated with bark, twigs, canella seeds, and star-anise.

David S. Brown Store (1875–1876)
8 Thomas Street, Manhattan
David S. Brown & Co. soap manufacturers commissioned J. Morgan Slade to design this whimsical structure. A favorite among architecture aficionados, it successfully incorporates many styles, including Venetian Gothic and Romanesque. Our reproduction uses tinted lemon leaves for the window arches, dried pomegranate for the roof over the door, and seagrass rope for columns. Magnolia leaves, pine cones, dried orange slices, corkscrew willow, cinnamon stick slices, juniper berries, whole peppercorns, red twig dogwood, and oak bark are used for ornamentation.

David S. Brown Store

Mrs. Josephine Schmid House (1895)
Fifth Avenue and 62nd Street, Manhattan
No longer in existence, this elaborate house was designed by Richard Morris Hunt, master of the chateau-style mansion. He created many similar mansions for various Astors and other wealthy New Yorkers in the late 19th century. This house was very long and narrow; only 25 feet wide on its Fifth Avenue side, it stretched to 100 feet on the 62nd street side. Our version has masonry of sand in tinted grout and reeds. The roof is made of oak leaves with dormers of dried pomegranate. The balcony railings are corkscrew hazel and grapevine tendrils. Poppy seed capsule tops, nigella seed capsules, and woodroses provide additional ornamentation.

United Nations Headquarters

South Street Seaport Historic District

Statue of Liberty

Statue of Liberty

Statue of Liberty (1871–1886)
Liberty Island, New York Harbor
A gift from the French to commemorate the centennial of the Declaration of Independence, the statue was designed by Frédéric Auguste Bartholdi and shipped to America in 350 pieces. The pedestal was designed by Richard Morris Hunt and paid for with private donations. From the ground to the tip of her torch the statue measures 305 feet. Our Lady Liberty's robes are made from palm fronds and grasses, her necklace from stalks of wheat, and her torch flame from a dried monarch flower inside a pomegranate half.

United Nations Headquarters (1947–1953)
United Nations Plaza, Manhattan
The U.N. complex consists of four buildings on 18 acres along the East River: the 39-story Secretariat building, the General Assembly, a Library, and a Conference Building. An international team of architects, led by Wallace K. Harrison and Le Corbusier, was assembled to design it. Our façade is made of sand and colored tile grout. The national flags are all different leaves.

South Street Seaport Historic District (ca. 1793–1811)
Fulton and South Streets, Manhattan
These warehouses and counting houses were built by Peter Schermerhorn, a merchant and ship owner, at a time when the area was New York's major shipping center. The corner building is made of white oak leaves, cinnamon sticks slices, walnut shells, bark, and honeysuckle. Sliced branches are used for the sign. The other buildings are made of eucalyptus seed pods and magnolia leaves, honeysuckle, cinnamon sticks, and pine cones for the roofs.

FIGS
NYC
42114

Philip and Maria Kleeberg House (1896–1898)
3 Riverside Drive, Manhattan
Charles Pierrepont Henry Gilbert, a prominent New York architect of the late 19th and early 20th centuries, designed this handsome house for the Kleebergs. Trained at the Ecole des Beaux-Arts in Paris, Gilbert's specialty was designing grand, chateau-style houses on Fifth Avenue for wealthy New York patrons, such as entrepreneur Frank Woolworth and investment bankers Felix Warburg and Otto Kahn. Our replica is made of hemlock cones, leaves, pine cone scales, acorn caps, pine bark nuggets, cattails, and grapevine. The roof is made of eucalyptus twigs, seeds, and leaves.

Sunnyside (1835)
Sunnyside Lane, Tarrytown
This romantic house overlooking the Hudson River was built by writer and diplomat Washington Irving as a retreat after decades of traveling abroad and exploring the American West. Originally a simple 18th-century stone cottage, Irving extensively enlarged and remodeled it in the 1830s with the help of his neighbor, George Harvey, a landscape painter. The façade of our version is made of sand in tinted grout. The roof, shutters, and door are made of leaves. Twigs frame the windows while cedrella pods provide the dormer eave decoration.

Samuel Trowbridge House

Samuel Trowbridge House (1902–1903)
123 East 70th Street, Manhattan
Samuel Beck Parkman Trowbridge's firm, Trowbridge & Livingston, designed many prominent commercial New York City buildings. Among their better-known commissions is the former B. Altman & Co. building (now part New York Public Library, part City University of New York), and the St. Regis Hotel. Trowbridge also designed his own house on 70th Street using a variety of decorative elements, such as iron balustrades, carved garlands, medallions, and elaborate pedimented window surrounds. Our version uses eucalyptus leaves, reeds, cinnamon sticks, bittersweet, canella berries, acorn caps, and anise seeds.

Eden Musée (ca. 1883)
55 West 23rd Street, Manhattan
Destroyed in 1916 and subsequently rebuilt at Coney Island, the Eden Musée featured "the usual retinue of freaks, midgets, fire eaters, sword swallowers, waxworks, [and] a Chamber of Horrors" as Luc Sante describes so well in his history of the seamy side of old New York, *Low Life*. Our version has a façade of bark and a roof made of oak leaves. The elaborate ornamentation is made of grapevine tendrils, nigella seed pods, dried kumquat slices, star-anise, corkscrew hazel, curls of birch bark, pine cone scales, and shelf fungus.

Samuel Trowbridge House

Eden Musée

Lighthouse (1872)
Lighthouse Park, Roosevelt Island
Legend has it that a patient at a nearby mental institution built this lighthouse at the tip of Roosevelt Island. An inscription on the lighthouse ashlar lends the story some credence: "This is the work was done by John McCarthy who built the lighthouse from the bottom to the top. All ye who do pass by may pray for his soul when he dies." Our replica is made from sand in tinted grout with pine and hickory bark accents. The details are fashioned from pear-shaped pods, grapevine tendrils, tinted magnolia leaves and moss.

New Amsterdam Theater (1902–1903)
214 West 42nd Street, Manhattan
This highly decorative, dramatic building helped turn Times Square into a glittering and festive theater district. The design by Herts & Tallant combined art nouveau style with Germanic medievalism and classicism. Our little theater is adorned with pepperberries, eucalyptus pods, cinnamon stick slices, acorn caps, pine cone scales, dried orange slices, bittersweet berries, star-anise, nigella seed pods, and pear-shaped pods.

Brownstone

Brownstone (ca. 1860)
Brooklyn
Brownstone is a soft, close-grained sandstone. Available locally, it was an inexpensive alternative to marble and limestone and quickly became popular in the mid-1800s for everything from Fifth Avenue mansions to humble rowhouses. Indeed, it was so popular the term became common for any rowhouse. A true brownstone, however, is an ordinary brick rowhouse with a brownstone front and a front stoop. Our front is sand in tinted grout; bark, twigs, corkscrew hazel, grapevine tendrils, cinnamon stick slices, star-anise, and moss are used for the architectural details.

Federal Style Rowhouse (ca. 1830)
153 Bleeker Street, Manhattan
During the 18th century, rowhouses gradually became larger and more elegant, although they still possessed the essential simplicity of the Federal style. Rather than the usual red brick, this house had a showy granite front and a columned doorway. True to Federal style otherwise, it had a front stoop, a steeply pitched roof, pedimented dormers, and was 25 feet wide. Our rowhouse has a façade of bark rectangles and branches with corkscrew hazel and grapevine tendril details. The roof is made of leaves with acorn caps on the chimneys.

Collectors' Club (1902)
22 East 35th Street, Manhattan
In real life, one of America's premier philatelic organizations occupies this elegant neo-Georgian, five-story brownstone in Manhattan. In 1902, McKim, Mead & White completely redesigned the building. With columns and portals along the front, our likeness is composed of broom and willow sticks, corkscrew hazel pieces, red-tinted magnolia leaves, grapevine, and acorn caps.

167 West 23rd Street (ca. 1898)
Manhattan
This building was originally a small residence in what subsequently became, for a brief period, a theater district. The house was transformed into an elaborately decorated cast-iron front building housing a store and offices. The rooftop tower is made of pressed metal, an extraordinary embellishment for a building of this size. Our replica is made of acorn caps, grapevine tendrils, reeds and twigs, cinnamon stick slices, and eucalyptus pods.

Federal Style Rowhouse

Brownstone

Collectors' Club

167 West 23rd Street

Aguilar Library

Luna Park Ticket Booth | Luna Park Arch | Luna Park Central Tower | Galveston Flood Building

Luna Park Ticket Booth, Arch, and Central Tower (1903)
Coney Island, Brooklyn
An Arabian theme was chosen for the 22-acre Luna Park — built to house the Trip to the Moon ride — and a multitude of spires and minarets were erected to achieve a fantastical and cheerful effect. Eventually, 1,200 towers, minarets and domes stood in the park imparting a fanciful other-worldliness. The Luna Park Ticket booth has been recreated with cinnamon sticks and grapevine tendrils. The Arch is made with gourds, eucalyptus root, basket jute, and hung with strands of canella berry light bulbs. Our version of the 125-foot high Central Tower has been fashioned from sand in tinted grout, tree branch slices, star-anise, walnut shells, dried orange slices, rose petals and various seeds and gourds.

Galveston Flood Building (1908)
Coney Island, Brooklyn
This building once housed a mechanical cyclorama show depicting the tidal surge associated with the deadly hurricane of September 8, 1900 that devastated Galveston. Here, sand in tinted grout is used for the façade with leaves of magnolia, moneta, and red-tinted eucalyptus. Stars of Texas, rendered in red-bud twigs, adorn each side of the building.

Galveston Flood Building

Claude A. Vissani House

Claude A. Vissani House (1889)
143 West 95th Street, Manhattan
This delightful house, a New York City landmark, is aptly described in the *AIA Guide to New York City* as "exuberant neo-Gothicism as if done in the Renaissance fashion by a student trained at the Ecole des Beaux-Arts." Our version has a façade of sand in tinted grout. The Gothic window arches are made of twigs and reeds. Acorn caps, various cones and seeds, canella berries, and cinnamon sticks form the architectural details.

Brooklyn Bridge (1867–1883)
Linking the boroughs of Manhattan and Brooklyn, this was one of the world's first steel-wire suspension bridges and its largest until 1903 when the Virginia Williamsburg Bridge — only 4 1/2 feet longer — was built. The bridge was designed by John Augustus Roebling, an engineer and inventor who died in an accident before construction began. His son, Colonel Washington A. Roebling, became the Chief Construction Engineer after his death. Our bridge is made of maple, oak, and elm bark and branches.

Bartow-Pell Mansion (1836–1842)
Pelham Bay Park, Bronx
The Pells were the Lords of the Manor of Pelham, a 50,000-acre tract chartered by King Charles II in 1666. The family's house was burned during the American Revolution because of their loyalist leanings and the estate reduced to 220 acres. Robert Bartow, a publisher and descendent of the Pell family, built the present stone, Greek Revival house. Our version's façade is made of hardwood barks while the second story porch is hemlock branches with grapevine tendril railings. The dentil molding beneath the honeysuckle eaves is made of pine cone scales.

overleaf
Enid A. Haupt Conservatory (1901–1902)
The New York Botanical Garden, Bronx
The Conservatory's design, by William R. Cobb of Lord & Burnham, was inspired by two major glasshouses in England: the Palm House of the Royal Botanic Gardens, Kew, and the 1851 Crystal Palace at Hyde Park. Conservatory architects were the first to use curvilinear steel frames and shaped glass. This airy structure defied the Victorian notion of absolute solidity, at times blending into the sky. Our replica is made of reeds and casting resin for the glass.

NEW YORK CENTRAL

ENID A. HAUPT CONSERVATOR
Sculpture at the Garden
Train Show

List of Photographs

Produced in association with
Ellen Bruzelius and
Patrick Filley Associates, Inc.

Design by Salsgiver Coveney Associates Inc.

Edited by Margaret Falk and Anne Steinbrook

Special thanks to Conservatory Manager Francisca Coelho and the Conservatory staff for their indispensable help in setting up and maintaining the show each year, and to Jennifer King for her valuable contribution.

Published by
Shop in the Garden Books
The New York Botanical Garden
Bronx, New York 10458

Printed in Hong Kong

ISBN 0-89327-959-5